AF674429

Mental Math

Workbook

Om KIDZ
An imprint of Om Books International

SHAPELY GUESS

Fill in the blanks to help the Mathesaur learn the names of these shapes.

This is a square.

This is a __________.

This is a __________.

This is a __________.

COLOUR ME QUICK

Follow the code and add some colour to the Mathesaur!

SUPER SIDES

Count the sides in each shape below and write the number within the circles.

SHAPELY COUNT

Look at the figure given below. Identify the shapes within and write their total number in the blank eggs below.

Number of triangles: 4

Number of squares:

Number of rectangles:

Number of circles:

FLOATY COUNT

Help the Mathesaur count the balloons. Fill in the correct numbers below.

Total number of balloons: 23

Total number of red balloons: ____________

Total number of blue balloons: ____________

Total number of yellow balloons: ____________

Total number of green balloons: ____________

ADDITION MISSION

Can you add the numbers on the rocks to help the Mathesaur learn addition?

- 13 + 2 = 15
- 20 + 4 =
- 18 + 6 =
- 9 + 8 =
- 12 + 6 =
- 27 + 2 =
- 17+ 5 =
- 11 + 15 =
- 23 + 8 =

ADDI-FUN!

Teach the Mathesaur how to add by solving the sums below.

$$\begin{array}{r} 2 \\ +\ 3 \\ \hline 5 \end{array}$$

$$\begin{array}{r} 1 \\ +\ 1 \\ \hline \end{array}$$

$$\begin{array}{r} 4 \\ +\ 3 \\ \hline \end{array}$$

$$\begin{array}{r} 1 \\ +\ 4 \\ \hline \end{array}$$

$$\begin{array}{r} 4 \\ +\ 1 \\ \hline \end{array}$$

$$\begin{array}{r} 5 \\ +\ 3 \\ \hline \end{array}$$

WHAT'S NEXT?

Fill in the missing numbers for the Mathesaur.

44	45	46	47
54	55		
64		66	
		76	77
84		86	

33	34	35	36
18			21
		83	84
	25	26	
		99	

20	30	40	50
60			90
10		30	
50	60		
	40	50	

BIRTHDAY MATH

This Mathesaur is having a birthday party!

Count the objects in the above picture and fill in the circles given below with the correct answer.

cupcakes	5	balloons		stars	
party hats		guests		gifts	

SUBTRACT-FUN

Count the fingers on the first hand. From this number, subtract the fingers on the second hand. What is the remaining number? Write it in the circles alongside.

MATH SWEET MATH

This Mathesaur is learning how to count!

Help the Mathesaur count and subtract the candies.
Write the correct answers on the blanks.

Blue candies	6	- Green candies	3	= 3
Red candies		- Yellow candies		= ______
Red candies		- Green candies		= ______
Blue candies		- Yellow candies		= ______

BIRTHDAY GIFT

This Mathesaur's aunt gave her $50 for her birthday. Which of these things can she buy with that money? Circle the correct pictures.

MATH FLOWERS

This Mathesaur is learning subtraction. Can you solve the sums on the flowers for her?

  5
- 3
―
  2

  9
- 2
―

  4
- 1
―

  7
- 3
―

 10
- 5
―

OFF TO SCHOOL

This Mathesaur has some extra school supplies. Count the objects in the picture and subtract them from the ones he already has.

The Mathesaur needs six pencils.
So, he has ___4___ extra pencils.

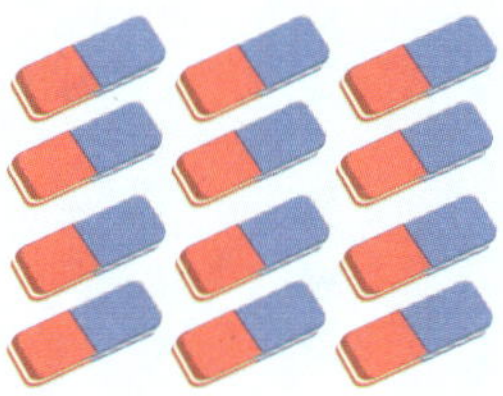

He needs three erasers.
So, he has ______ extra erasers.

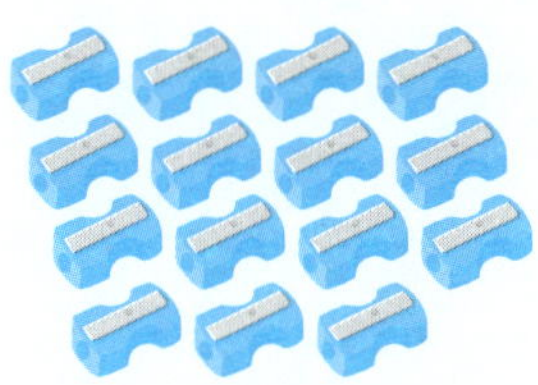

He needs seven sharpeners.
So, he has ______ extra sharpeners.

He needs five pens.
So, he has ______ extra pens.

MINUS MENACE

Choose the correct answer from the box and help the Mathesaur learn subtraction.

				Answer box
9 -	6	= 3		3
7 -		= 4		4
2 -		= 0		3
8 -		= 3		2
6 -		= 2		5
4 -		= 1		6

MONEY MATH

Help the Mathesaur add money by writing the correct answer below.

TWENTY DOLLARS + TEN DOLLARS = $30

FIFTY DOLLARS + ONE HUNDRED DOLLARS = ______

FIVE DOLLARS + ONE DOLLAR = ______

TWENTY DOLLARS + FIVE DOLLARS

+ FIVE DOLLARS = ______

NUMBER CIRCUS

Circle the odd numbers and underline the even numbers for the Mathesaur.

1	2	3	4
5	6	7	8
9	10	11	12
13	14	15	16
17	18	19	20
21	22	23	24
25	26	27	
28	29	30	

WHAT'S HEAVIER?

This Mathesaur is trying to rank the objects according to their weight. Help him by numbering the objects. 1 is for heavy objects and 2 is for light objects.

WHO'S THE HEAVIEST?

Write 1, 2 and 3 to rank the objects from lightest to heaviest.

MEASURING MATH

Measure the pencils against the ruler and write their lengths in centimetres (cm).

____2____cms

________cms

________cms

________cms

________cms

HEIGHT LADDER

List the objects according to their height on the ladder below. Begin with the smallest and move down the ladder to the biggest object.

LENGTHY LABELS

Can you label these objects according to their length?

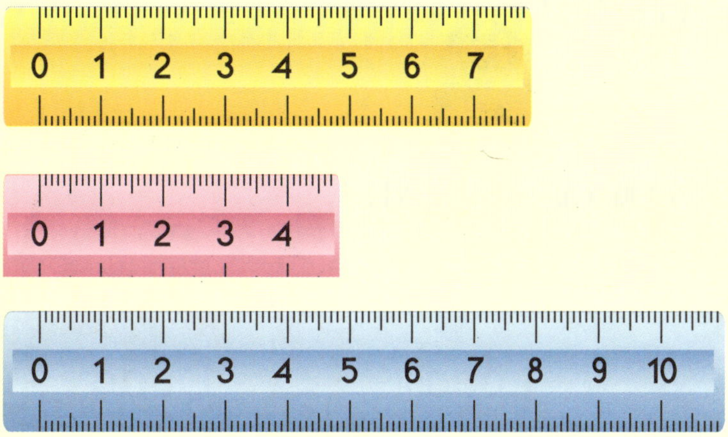

longer

long

longest

TABLE TROUBLE

Look at the table and answer the questions.

	Pogo	Lopa	Dodo
Weight	15 kg	20 kg	30 kg
Height	2 feet 8 inches	3 feet 5 inches	3 feet

1. What is Dodo's height? ____3 feet____
2. What is Pogo's weight? ______________
3. Who is the shortest? ______________
4. Who is the heaviest? ______________

TIME IT

Help the Mathesaur read the time. Draw the hands of the clock according to the time written below.

5:00

11:30

9:30

6:00

12:00

1:30

SUPER SORTER

This boy is getting ready for school. Can you number his pictures in the correct order? If you need help, you can look at the clocks!

SMALL TO BIG

Rearrange the numbers in an ascending order.

48	50	23	23	48	50
27	19	31			
64	87	78			
52	35	42			
30	21	19			
84	92	73			
77	68	89			
14	35	29			

DAILY MATCH

Match the days to the correct numbers.

JIGSAW PUZZLE

Tick the piece that fits in the puzzle correctly.

JIGSAW NUMBERS

If you add the numbers in the circles, their sum is the number in the square between them. Can you solve this puzzle?

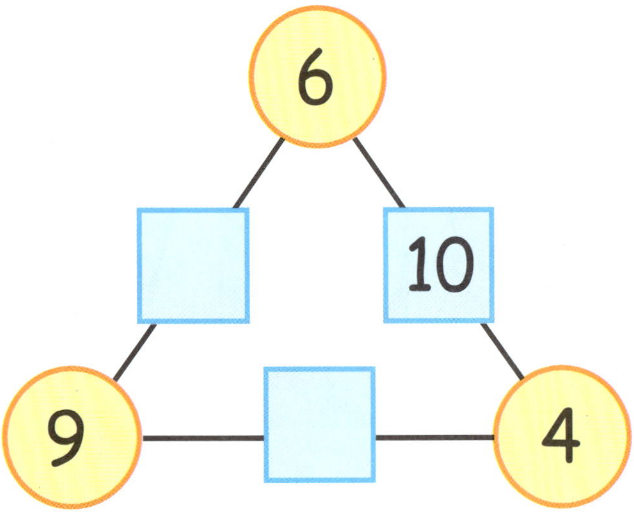

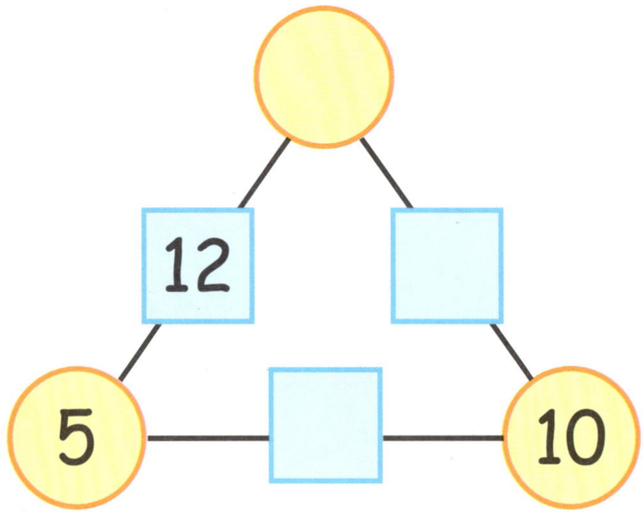

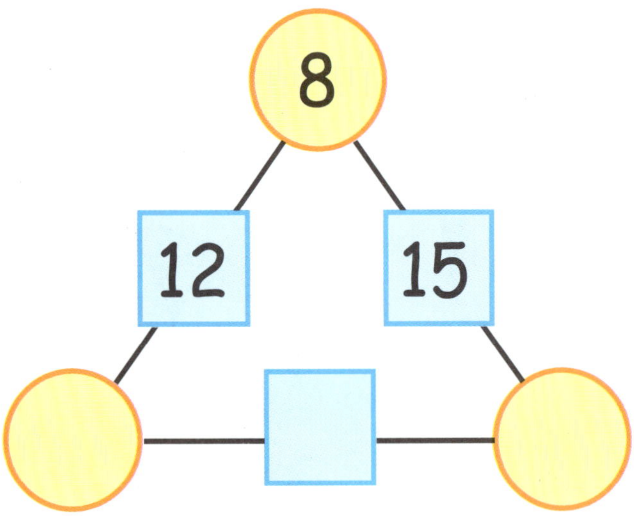

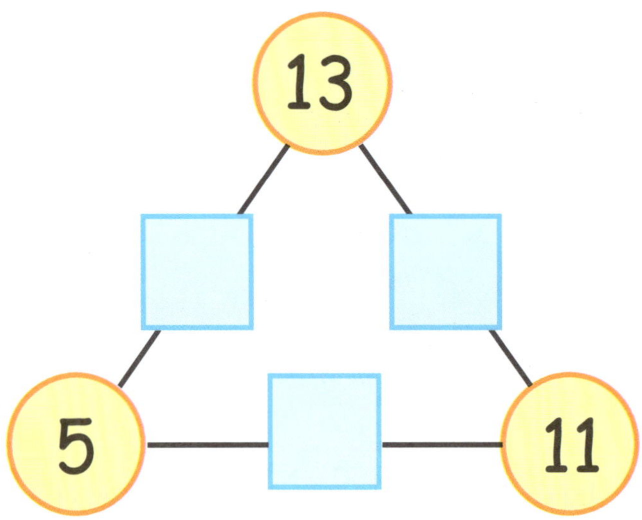

ANSWER KEY

Page 2

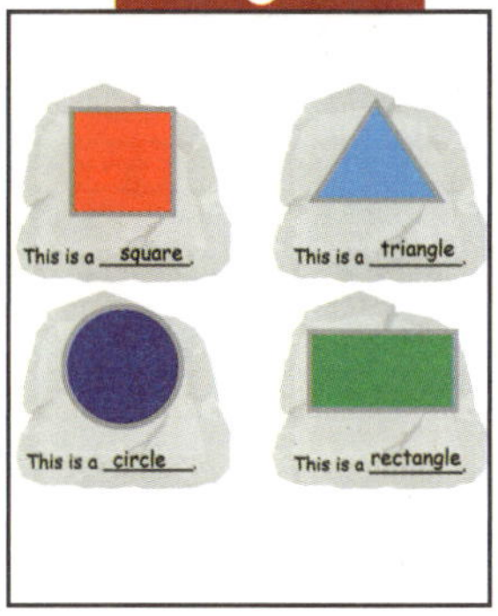

Page 3

Page 4

Page 5

Page 6

Page 7

Page 8

Page 9

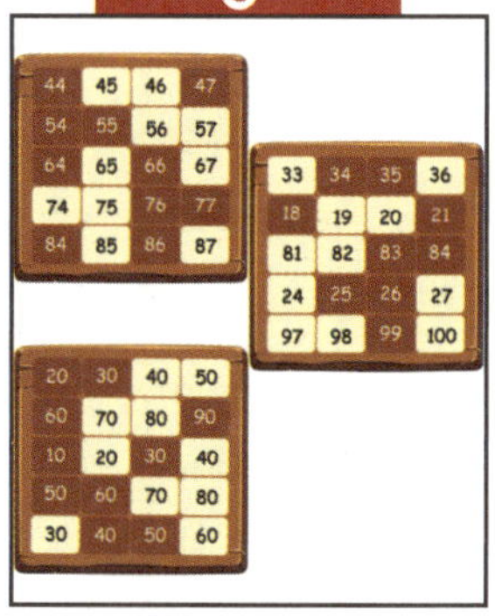

Page 10

Page 11

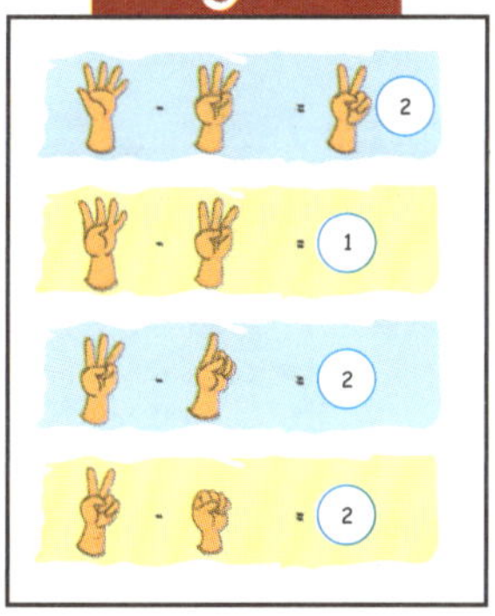

Page 12

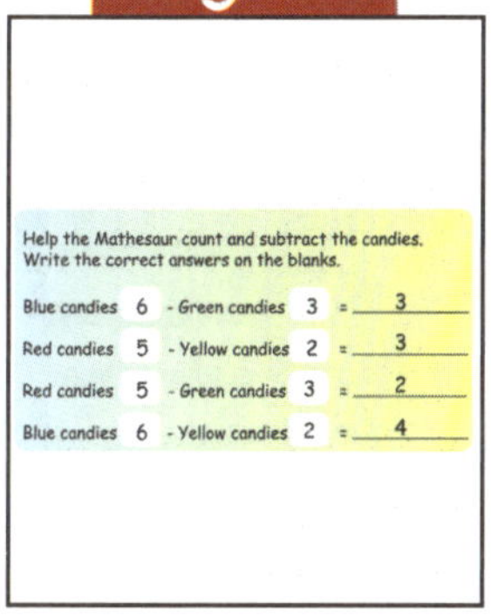

Page 13

Page 14

Page 15

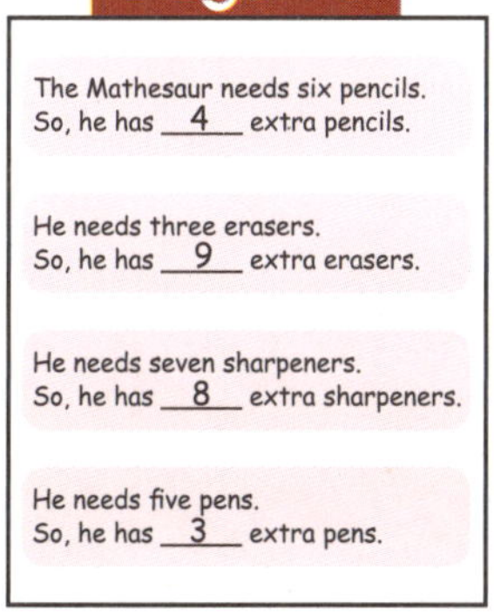

Page 16

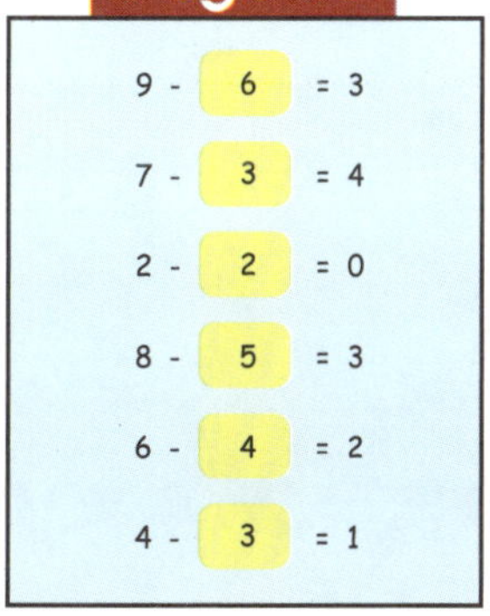

Page 17

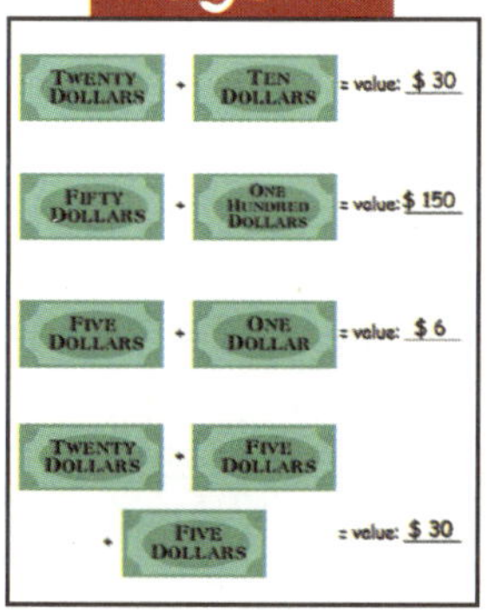

ANSWER KEY

Page 18

Page 19

Page 20

Page 21

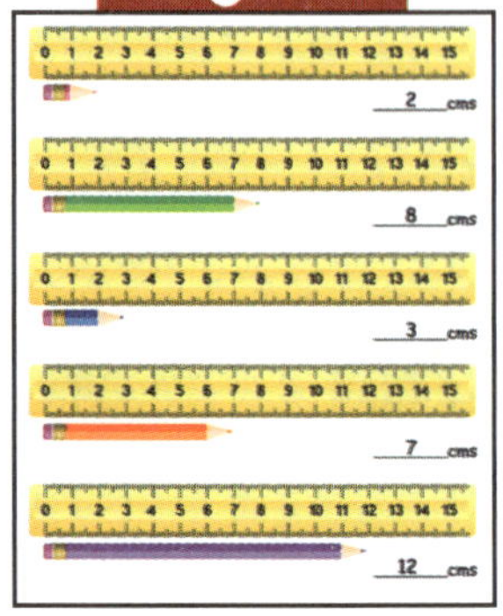

Page 22

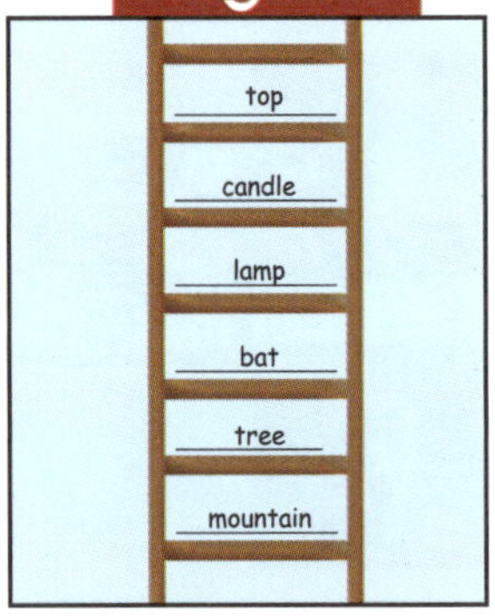

Page 23

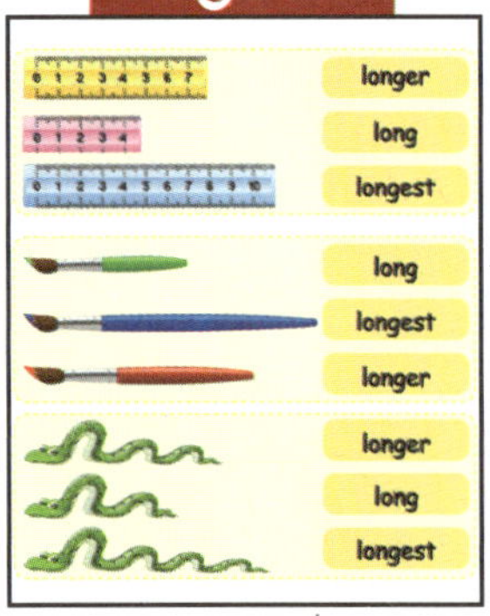

Page 24

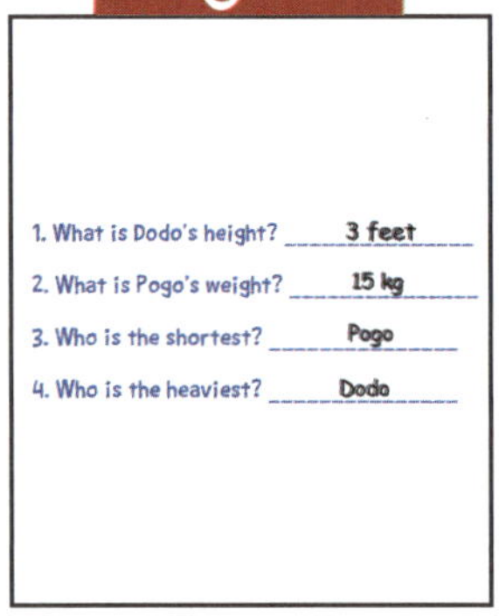
1. What is Dodo's height? 3 feet
2. What is Pogo's weight? 15 kg
3. Who is the shortest? Pogo
4. Who is the heaviest? Dodo

Page 25

Page 26

Page 27

48	50	23	23	48	50
27	19	31	19	27	31
64	87	78	64	78	87
52	35	42	35	42	52
30	21	19	19	21	30
84	92	73	73	84	92
77	68	89	68	77	89
14	35	29	14	29	35

Page 28

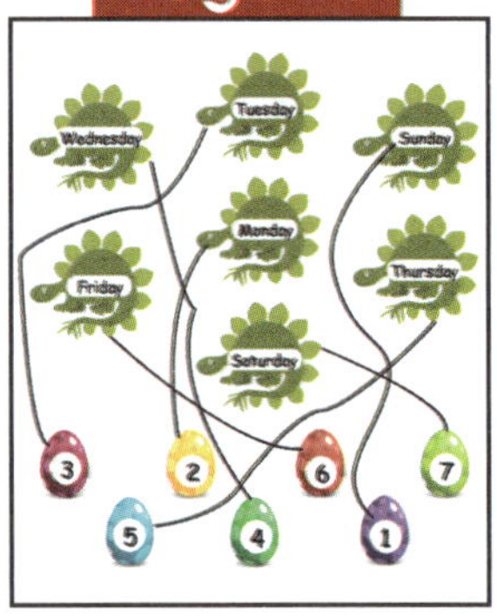

Page 29

Page 30

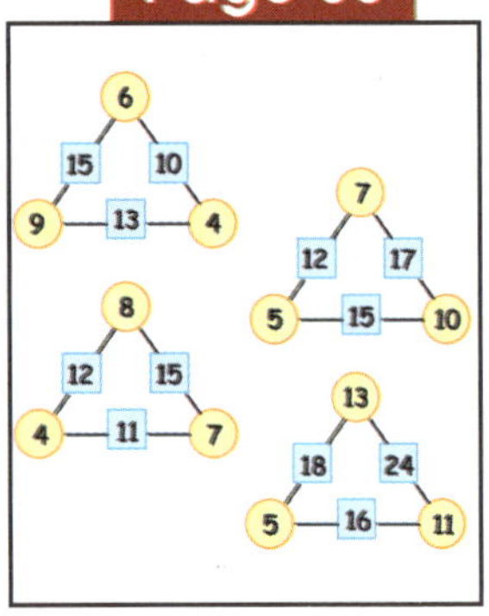